Frigate USS Clark

ONE NAME, TWO SHIPS and two admirals

It is very rare with navies from all nationalities to have had two ships with the same name that were not named after the same person. It goes without saying that ships with the same name cannot serve with the same navy at the same time. That could lead to a lot of confusion. Normally ships are named after for instance naval heroes. After a number of years that ship is struck from the list so the name disappears. Sometimes, only many years later, but also in a continuous line a new ship is given the same name. If you look at the names given with different navies there is mostly quite a long time between two ships that carry the same name. This is also the case with the US Navy ships named *Clark*.

The then king Willem I decided that there would always be a ship with the name Van Speyk in the Dutch navy and this has been the case from that moment on. Sometimes a *Van Speijk* is struck from the list and immediately succeeded by a ship that allotted that name. More often an existing ship is rechristened.
As indicated above two ships in the US Navy carried the name *Clark*. These ships did not only sail in two different eras with totally different global situations but it is most particular that the ships were not named after the same person. Of course this book is about the second ship of this name but it is interesting to briefly discuss the first *USS Clark*.

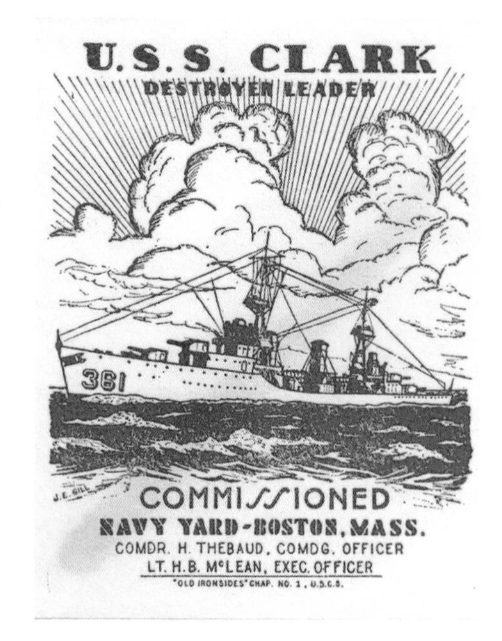

In the 1930s, the United States Navy built two classes of flotilla leaders, the Porter class, and the Somers class. Due to the regulations of the London Naval Treaty, these 13 ships had a displacement of 1,850 tons, compared to the 1,500 tons of a "standard" destroyer. When the treaty ended with the outbreak of World War II in Europe in 1939, the much larger Fletcher class was constructed, making the differentiation irrelevant.

The national ensign consists of thirteen equal horizontal stripes of red (top and bottom) alternating with white, with a blue rectangle in the canton bearing fifty small, white, five-pointed stars. The stars represent the fifty states of the United States of America and the 13 stripes represent the thirteen British colonies that declared independence and became the first states in the Union.

The first *Clark* was commissioned in 1936 and the second *Clark* in 1980. There was no ship with this name between 1946 and 1980.
In special cases there is always a ship with a particular name active with a navy.

A clear example is the name Van Speyk. He sacrificed himself during the Belgian insurgence of 1830 by blowing up himself and his ship. This to avoid that his ship and crew would fall in the hands of the Belgian insurgents when Belgium was still part of the Kingdom of the Netherlands.

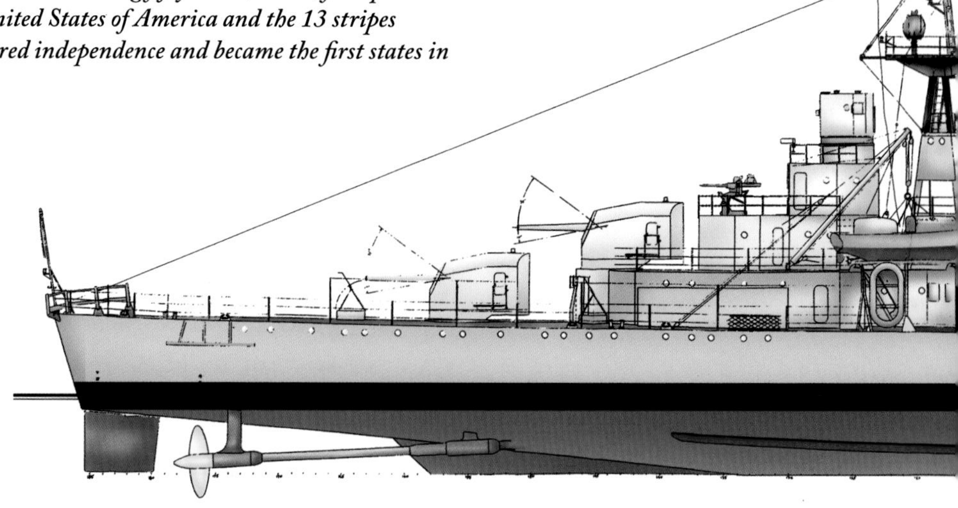

This USS Clark (DD-361) was a Porter-class destroyer of the US Navy.

THE FIRST USS CLARK

This USS *Clark* (DD-361) was a *Porter*-class destroyer of the US Navy.

This ship was named after Rear Admiral Charles E. Clark. He was born in 1843 in Bradford and graduated from the Naval Academy in 1863. During the Civil War he commanded the Ossipee during the battle of Mobile Bay. At the outbreak of the Spanish-American war in 1898 Captain Clark commanded the USS *Oregon*. He managed to arrive at Cuba with his ship after a stormy rounding of Cape Horn on time to be part of the destruction of the Spanish fleet. This great example of courage and perseverance led him to be promoted to Rear Admiral. He passed away on October 1, 1922 in Long Beach California.

The cause of the Spanish War was the Cuban request for US support to achieve independence. The Spanish would suffer huge numbers of casualties and it would confirm the name and fame of commodore George Dewy. His fleet destroyed the Spanish fleet in Manilla Bay in a short fight in which the Americans suffered only 8 wounded. Dewy was immediately promoted to Rear Admiral by president McKinley. The war with Spain was not over yet. A second sea battle took place after 17.000 men landed on Cuba in June under the command of General W.R.Shafter. They attacked the Spanish fleet from land. Within three hours the Spanish fleet under the command of admiral Cervera was completely destroyed. The Spanish suffered over 600 casualties and more than 1700 Spanish were taken prisoner. The Americans suffered only one casualty and one wounded.

Rear Admiral Charles Edgar Clark (10 August 1843 - 1 October 1922).

The large prewar 'leaders' were almost cruiser-like, as the Clark shows, with her tripod masts and enclosed twin 5 inch guns.

USS CLARK DD-361

DD-361 was launched on 15 October 1935 at the Bethlehem Shipbuilding corporation in Quincy Massachusetts and baptized by Mrs S.Robinson. She received her commission on May 20 1936 under the command of Commander H.Thebaud. Before World War Two, USS *Clark* served on the Atlantic coast, the Caribbean and from Pearl Harbor. That became her home port from April first 1940. From Pearl Harbor she sailed to Samoa, the Fiji Islands and Australia. At the outbreak of war she was in San Diego California for major maintenance work. On December 27 she sailed from the West Coast to escort two convoys to Pearl Harbor. After that she carried out anti-submarine patrols near Pago Pago. In February and March of 1942 she was part of a carrier taskforce for attacks on New Guinea. From April she escorted a large number of convoys from Noumea that supplied fuel, parts etc. for Carrier Task Forces. In December 1942 she sailed to Balboa to serve as the flagship of the Commander Southeast Pacific Force. Until 1944 she patrolled from a number of American ports. Between early September 1944 and mid April 1945 she docked for major maintenance. Until the end of the war she was part of six trans-Atlantic convoys between the USA and the United Kingdom. She arrived in Philadelphia on June 15 1945 where she was decommissioned on October 23. She was broken up soon after that in March 1946. For her efforts during the Second World War she was awarded two Battle Stars.

A rubber stamp on the ship's mail.

Technical data
USS *Clark* DD -361

Class	*Porter* class destroyer
Displacement	1.850 tons; 2.597 tons fully loaded
Length	116 m. (381 ft.)
Beam	11 m. (36 ft.)
Draught	3 m. (10 ft.)
Propulsion	50.000 shp. Two geared turbines at 2 propellers Babcock & Wilcox boilers
Speed	37 knots
Range	6.500 nautical miles. at 12 knots
Complement	194 officers and enlisted

The *Porter*-class comprised eight ships in total. Only one of these ships, the name ship of the class USS *Porter* DD-356 was lost on 26 October 1942. She was torpedoed by the Japanese *I-21* in the Battle of Santa Cruz Islands, where she was a screening unit of TF61 (with the carriers *Enterprise* and *Hornet*).
Porter was torpedoed, and, after the crew had abandoned ship, sunk by gunfire from *Shaw* (DD-373). Although many believe she was torpedoed by Japanese submarine *I-21*, Japanese records don't support this. More likely, an errant torpedo from a ditching US TBF *Avenger* hit *Porter* and caused the fatal damage.

Technical data

Armament

As built	Guns: 1 x MK 33 Gun Fire Control System; 8 x 5 inch (4 x 2) 127 mm./38 cal; 8 x 1.1 inch 28 mm (2 x 4) AA; 8 x 21 inch 533 mm torpedo tubes (2 x 4); 2 x Depth Charge stern racks
In 1945	1 x MK 27 Gun Fire Control System; 8 x 5 inch (4 x 2) 127 mm./38 cal; 4 x MK 51 Gun Directors; 12 x Bofors 40 mm. (2 x 4 and 3 x 2); 6 x Oerlikons 20 mm. (6 x 1); torpedo tubes (2 x 4); 2 x Depth Charge stern racks

Abbreviations

The United States Navy normally indicates the type of ship before the pennant number. DD means Destroyer. The list of abbreviations runs as follows.

BB	Battleship
CA	Heavy Cruiser
CL	Light Cruiser
CLAA	Light Cruiser Air Defence
CV	Airplane Carrier
CVL	Light Airplane Carrier
CVE	Escort Airplane Carrier
SS	Submarine
DD	Destroyer
DE	Escort Destroyer
LCI	Landing Craft Infantry
LCM	Landing Craft Mechanized
LCT	Landing Craft Tanks
PC	Submarine Chaser
PT	Torpedo Boat

Besides these class indicators we also find the character N for Nuclear. For ship propelled by Nuclear Energy. For example; CVN USS *Enterprise*.

USS Clark entering port in 1941. The ships of the Porter- and Somers class were quite badly overweighted. During WW2 they were rebuilt with dual-purpose guns and their Mk 35 directors refurbished.

Other ships of this class:
- DD-357 USS *Selfridge*;
- DD-358 USS *Mc Dougal*;
- DD-359 USS *Winslow*;
- DD-360 USS *Phelps*;
- DD-362 USS *Moffett* and
- DD-363 USS *Balch*.

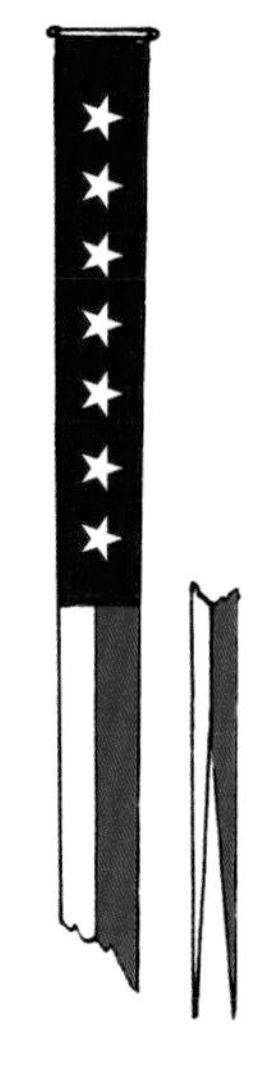

In the US Navy, the commissioning pennant is flown day and night at the loftiest point on the aftermost mast, from the moment the ship is commissioned until the moment she is decommissioned.

Admiral Joseph J. Clark

Joseph James Clark was born on 12 November 1893 in Chelsea Oklahoma. He was of Cherokee Indian descent. In 1918 he was the first native American to graduate from the United States Naval Academy. This is where he was nicknamed 'Jocko'. By the end of World War One he served on the cruiser USS *North Carolina* and witnessed the Atlantic Convoys. In 1921 he took command of the destroyer USS *Brooks*. In 1925 he started in a way a new career as pilot with the Navy flying corps. In this way he started a career in which he stood at the cradle of the developments of the Navy flying corps and the use of airplane carriers in war. Knowledge and competence that would be of great value and that would make him one of the most skilful commanders of the carrier fleet of the US Navy and commander of larger fleets.

At the outbreak of the Second World War he served as Executive officer on USS *Yorktown* but he became commander of USS *Suwanee* CVL-27 on the Atlantic. When the new carrier USS Y*orktown* CV-10 was commissioned, Joseph Clark became its first commander. During the carrier raids in 1943 the *Yorktown* was the flagship of admiral Pownall. This made it possible for him to gain extensive experience in the execution of attack plans using carriers and planes. In 1944 he was promoted to Rear Admiral and the command of Task Force 58. He was in command during the battle for the Marianas and the battle of the Philippine Sea. His flagship was USS *Hornet* CV-12. On the second day of the battle when the planes returned to the carriers at sundown he ordered to lit the landing lights of the carriers so that the pilots could find their ship and most of them landed safely.

During the Korean War he was commander of Fast Carrier Task Force (TF 77) and later Vice Admiral of the Seventh Fleet.

He retired as a full admiral on December 1 1953. He passed away on 13 July 1973 in St. Albans New York and was buried at Arlington National Cemetery.
He was awarded the following distinctions:
Navy Cross, Distinguished Service Medal, Legion d'Honneur (France) Legion of Merit, the Navy Commendation Medal en de Korean Order of Military Merit.

The Union Jack

The US Union Jack consists of the blue canton of the ensign.

Jacks are displayed at the jackstaff, a pole mounted on the bow of the ship, on ships in commission or in service. They are never flown ashore. The size of the jack is always the same as the size of the canton of the ensign at the flagstaff. The jack is displayed only during the hours between 8:00 a.m. and sunset, when the ship is not underway and the ensign is flying on the flagstaff. If the ensign is half-masted, so is the jack.

The "first Navy Jack" was one of a number of rattlesnake flags popular in the Continental Navy as well as throughout the colonies during the American Revolution. The symbolism is of a deadly animal that strikes only after giving fair warning.
On May 31, 2002, Secretary of the Navy directed that "the first Navy jack will be displayed on board all US Navy ships in lieu of the Union Jack, in accordance with Navy Regulations" for the duration of the global war on terrorism. Display of the striped jack throughout the fleet actually began at morning colors on September 11, 2002, first anniversary of the terrorist attacks on New York and the Pentagon.

The Perry-frigate USS Ford (FFG-54) in August 1986. The considerable separation between the MK 92 radome and the STIR tracker-illuminator farther aft is a survivability feature, since it is unlikely that a single hit would destroy both missile-guidance channels.

HISTORIC BACKGROUND TO THE DEVELOPMENT OF THE OLIVER HAZARD PERRY CLASS

The US Navy had a large number of ships at its disposal at the moment of the Japanese capitulation in 1945. Many were mothballed, orders for new ships were cancelled and a large number of ships sold to allied states under favourable conditions as part of mutual defence aid programmes. However, a large fleet was kept on active duty because of the deteriorating relation with the USSR. This would result in the so called Cold War. The intention was to limit the expanding power of the Soviet Union as much as possible. This resulted in an arms race and the development of the strategic nuclear arms. In 1972 president Nixon signed the first treaty between both nations to limit the number of strategic nuclear weapons. This treaty is known as SALT 1 (Strategic Arms Limitation Treaty). The effect was that there would not be a clear vision on the further development of the armed forces during the following administrations. President Carter did not only cut back the defence budget. He also cancelled a lot of the ships on the list to be built that was drafted during the Ford administration. The building programme was cut back from 157 ships to be built to 67. Thus in spite of the fact that the USSR budget rose and that they were building a strong fleet. The number of ships of the US Navy went from over 900 ships in 1970 to approximately 460 in 1978.

From 1979 political developments in the world gradually reversed the downward spiral of the US defence force and thereby also of the US Navy.
Problems caused by the countries in the Middle East, bordering Hormuz Strait, Persian Gulf and the Arabic Sea showed

the Americans that they were dependent on transport through these areas for one third of their oil consumption. The USSR were extending their influence in this region through the occupation of Afghanistan as well as in other parts of the world. This showed that relying on nuclear arms, such as those on nuclear submarines, to maintain the safety of the USA and its allies was a misguided idea in case of conflicts like these.

A new kind of Cold War was developing. That necessitated a change in thought by the American Congress, the Senate and the American people. The first years of the Reagan administration showed this change. Reagan made this clear in one of his speeches. *'Freedom to use the seas is our nation's lifeblood. For that reason our Navy is designed to keep the sea lanes open worldwide.'*
Reagan's vision of a strong navy resulted in the so called '600-vessels-navy. Reagan's arguments were strongly and positively reinforced by the Falkland Islands war between Great Britain and Argentina. This conflict showed the major importance of a strong expeditionary force based on a strong navy in all its aspects for the defence and safeguarding of a nation's interests.

Total number of ships in the US Navy

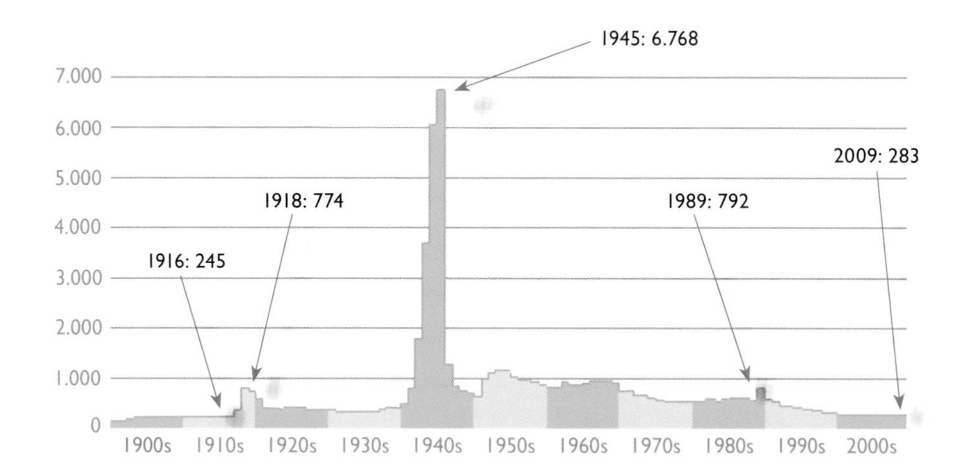

1200 US marines were stationed in Lebanon in September 1982 as part of an international peacekeeping force. When these forces were subjected to heavy fire from the militias in the hills surrounding Beirut airport the big guns of the

USS Rentz (FFG 46). On 5 November 1986, the frigate was part of an historic visit to Qingdao (Tsing Tao;) China—the first US Naval visit to China since 1949.

Battle Star

The Army, Air Force, Navy and other uniformed services of the United States award 'Service Stars' when someone took part in a particular operation. This may be a Bronze Star or a Silver Star. A Silver star is valued at five Bronze Stars. On the uniform one wears a 'ribbon' with or without stars. A ribbon without star means participation in one campaign. In the past for instance the campaigns in Korea, Vietnam, Kosovo and Iraq. Nowadays stars are also awarded to people who engaged in the War on Terrorism.

Historically during World War Two and the Korean War recommendations for a distinction, the so called Battle Stars were also awarded to ships of the US Navy. These were awarded for heroic conduct in a sea battle or severe damage sustained during a battle. These can be compared to the Battle Effectiveness Award (Battle E) that were awarded during the Vietnam War. These replaced the Battle Star. This is still common usage.

Nowadays 'ribbons' awarded for participation in peace time operations are often painted somewhere on the superstructure of the ship.

Command Excellence Awards are painted and displayed on the port and starboard side of the bulwark, aft of the Battle "E".

Black "E" - Maritime Warfare Excellence Award
Red "E" - Engineering/Survivability Excellence Award
Green "E" - Command & Control Excellence Award
Green "H" - Health and Wellness (Medical) Excellence Award
Blue "E" - Logistics Management Excellence Award
Yellow "E" - Commander, Naval Surface Forces (CNSF) Ship Safety Award
Purple "E" - Efficiency Excellence Award

battleship USS *New Jersey*, brought back into active duty from the mothball fleet, relieved them from these attacks.

As a result of these developments a huge building programme came into being. As well as a renovation of the fleet concerning weapon systems, planes and ships. In addition the *Spruance*- and *Kidd*-class destroyers were added to the fleet as well as the *Perry*-class frigates and the Aegis cruisers. In 1983 the fleet counted 490 vessels and 76 ships were being built. The *Oliver Hazard Perry*-class was the largest series of ships of a single class built after World War Two.

The ribbons and excellence awards on the superstructure of USS Clark.

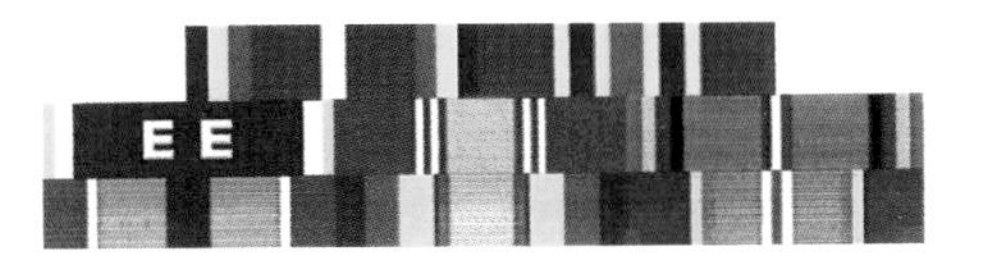

Left: Precedence of awards is from top to bottom, left to right.

Top row: *CG Unit Commendation - Navy Meritorius Unit Commendation*
Second row: *Navy Battle 'E' ribbon (2) - National Defense Service Medal - Armed Forces Expeditionary Medal.*
Third row: *Humanitarian Service Medal - Navy Service Deployment Ribbon - CG Special Operations Service Ribbon.*

USS Oliver Hazard Perry, the first ship of class seen here in the Wieland Canal (Canada) in June 1996. Much of the boxiness of her appearance can be blamed on a decision half-way through her design to add a second helicopter, requiring a much wider hangar. Later units had enlarged fantails and modified hangars to accommodate a larger helicopter.

THE OLIVER HAZARD PERRY-CLASS

Knox-class as completed.

Ships of this class were built in the seventies to replace the obsolete ships from World War Two and the *Knox*-class frigates from the sixties. An important design condition was that they had to be cheap enough to be built in large numbers. They were designed as 'general purpose escort vessels' that could be deployed for a wide number of tasks such as the protection of amphibious landings and the escort of convoys. Also the FFG's would be stationed in a Carrier Battle Group, screening capital units (HVU's) and the main body (Carrier in station zero).

These ships are also capable of operating independently against drugs trafficking. Inspection missions and combined exercises/operations with other navies . In total 71 ships of this type were built. During this period that extended over a fair number of years many modifications were made.
55 ships were built in the US:
- 51 for the US Navy and
- four for the Australian Navy.

A further eight ships were built for Taiwan by the China Shipbuilding Yard in Kaohsiung Taiwan, denoted as the *Cheng Kung*-class. Six ships for the Spanish Navy by Bazan in Ferrol denoted as the *Santa Maria*-class and two ships for the Australian Navy by Amecon in Williamstown denoted as the *Adelaide*-class. Former ships of the US Navy were sold or presented to the navies of Bahrain, Egypt, Pakistan, Poland and Turkey.

The ships were built in two versions. One 'short hull-445 feet' (136 metres) and the 'long hull-453 feet' (138 metres) . The short hull ships carry the SH-2 Seasprite LAMPS (Light Airborne Multiple System) helicopter . The other ships have the more powerful SH-60 Seahawk LAMPS III. The ships with pennant numbers 8, 29, 32 and 33 were designed and built as short hull. They were later modified to long hull. Ships with pennant numbers 33-61 were all built as long hull. They have a steel hull but the superstructure was designed in aluminium. There were doubts about this combination with the experts such as the Engineer in Chief of the British Royal Corps of Naval Constructors Admiral R.J. Daniels, during a discussion with Admiral R.C. Gooding, US Chief of the Bureau of Ships, about the variable pitch propeller. In effect a number of problems occurred with the superstructure. A 40 feet tear in USS *Duncan* being the most striking. These problems were solved as the building programme progressed.

Ships of this class are often referred to as "FFG-7" (pronounced Fig-Seven) after the lead ship, USS *Oliver Hazard Perry* (FFG-7).

US built Oliver Hazard Perry frigates

Name	Pennant	Builder	Service	Fate
Oliver Hazard Perry	FFG-7	Bath Iron Works	1977–1997	Disposed of by scrapping, dismantling, 21 April 2006
McInerney	FFG-8	Bath Iron Works	1979–2010	Transferred to Pakistan as PNS *Alamgir* (F-260)
Wadsworth	FFG-9	Todd Pacific Shipyards (Todd), San Pedro	1978–2002	Transferred to Poland as ORP *Gen. T. Kościuszko* (273)
Duncan	FFG-10	Todd, Seattle	1980–1994	Transferred to Turkey as a parts hulk
Clark	FFG-11	Bath Iron Works	1980–2000	Transferred to Poland as ORP *Gen. K. Pułaski* (272)
George Philip	FFG-12	Todd, San Pedro	1980–2003	Decommissioned, to be disposed of, 24 May 2004.
Samuel Eliot Morison	FFG-13	Bath Iron Works	1980–2002	Transferred to Turkey as TCG *Gokova* (F 496)
John H. Sides	FFG-14	Todd, San Pedro	1981–2003	Decommissioned, to be disposed of, 24 May 2004.
Estocin	FFG-15	Bath Iron Works	1981–2003	Transferred to Turkey as TCG *Goksu* (F 497)
Clifton Sprague	FFG-16	Bath Iron Works	1981–1995	Transferred to Turkey as TCG *Gaziantep* (F 490)
built for Australia as HMAS Adelaide	FFG-17	Todd, Seattle	1980–2008	Disposed, sunk as diving & fishing reef, 13 April 2011
built for Australia as HMAS Canberra	FFG-18	Todd, Seattle	1981–2005	Disposed, sunk as diving & fishing reef, 4 October 2009
John A. Moore	FFG-19	Todd, San Pedro	1981–2000	Transferred to Turkey as TCG *Gediz* (F 495)
Antrim	FFG-20	Todd, Seattle	1981–1996	Transferred to Turkey as TCG *Giresun* (F 491)
Flatley	FFG-21	Bath Iron Works	1981–1996	Transferred to Turkey as TCG *Gemlik* (F 492))
Fahrion	FFG-22	Todd, Seattle	1982–1998	Transferred to Egypt as *Sharm El-Sheik* (F 901)

Lewis B. Puller	FFG-23	Todd, San Pedro	1982–1998	Transferred to Egypt as *Toushka* (F 906)
Jack Williams	FFG-24	Bath Iron Works	1981–1996	Transferred to Bahrain as RBNS *Sabha* (FFG-90)
Copeland	FFG-25	Todd, San Pedro	1982–1996	Transferred to Egypt as *Mubarak* (F 911), renamed *Alexandria* in 2011
Gallery	FFG-26	Bath Iron Works	1981–1996	Transferred to Egypt as *Taba* (F 916)
Mahlon S. Tisdale	FFG-27	Todd, San Pedro	1982–1996	Transferred to Turkey as TCG *Gokceada* (F 494)
Boone	FFG-28	Todd, Seattle	1982–2012	Decommissioned 23 February 2012
Stephen W. Groves	FFG-29	Bath Iron Works	1982–2012	Decommissioned 24 February 2012
Reid	FFG-30	Todd, San Pedro	1983–1998	Transferred to Turkey as TCG *Gelibolu* (F 493)
Stark	FFG-31	Todd, Seattle	1982–1999	Disposed of by scrapping, dismantling, 21 June 2006
John L. Hall	FFG-32	Bath Iron Works	1982–2012	Decommissioned 9 March 2012
Jarrett	FFG-33	Todd, San Pedro	1983–2011	Decommissioned, held for future foreign military sale
Aubrey Fitch	FFG-34	Bath Iron Works	1982–1997	Disposed of by scrapping, dismantling, 19 May 2005
built for Australia as *HMAS Sydney*	FFG-35	Todd, Seattle	1983-	
Underwood	FFG-36	Bath Iron Works	1983-2013	Decommissioned 8 March 2013
Crommelin	FFG-37	Todd, Seattle	1983-2012	Decommissioned 26 October 2012
Curts	FFG-38	Todd, San Pedro	1983-2013	Decommissioned 25 January 2013.
Doyle	FFG-39	Bath Iron Works	1983-2011	Decommissioned 29 July 2011
Halyburton	FFG-40	Todd, Seattle	1983-2014	Decommissioned 8 September 2014
McClusky	FFG-41	Todd, San Pedro	1983-	
Klakring	FFG-42	Bath Iron Works	1983–2013	Decommissioned 22 March 2013
Thach	FFG-43	Todd, San Pedro	1984-2013	Decommissioned 1 November 2013
built for Australia as *HMAS Darwin*	FFG-44	Todd, Seattle	1984-	
De Wert	FFG-45	Bath Iron Works	1983-2014	Decommissioned 4 April 2014
Rentz	FFG-46	Todd, San Pedro	1984-2014	Decommissioned 23 May 2014
Nicholas	FFG-47	Bath Iron Works	1984-2014	Decommissioned 17 March 2014
Vandegrift	FFG-48	Todd, Seattle	1984-	
Robert G. Bradley	FFG-49	Bath Iron Works	1984-2014	Decommissioned 28 March 2014
Taylor	FFG-50	Bath Iron Works	1984-	
Gary	FFG-51	Todd, San Pedro	1984-	
Carr	FFG-52	Todd, Seattle	1985-2013	Decommissioned 13 March 2013
Hawes	FFG-53	Bath Iron Works	1985–2010	Decommissioned 10 December 2010
Ford	FFG-54	Todd, San Pedro	1985-2013	Decommissioned 31 October 2013
Elrod	FFG-55	Bath Iron Works	1985-	
Simpson	FFG-56	Bath Iron Works	1985-	
Reuben James	FFG-57	Todd, San Pedro	1986-2013	Decommissioned 30 August 2013
Samuel B. Roberts	FFG-58	Bath Iron Works	1986-	
Kauffman	FFG-59	Bath Iron Works	1987-	
Rodney M. Davis	FFG-60	Todd, San Pedro	1987-	
Ingraham	FFG-61	Todd, San Pedro	1989-	

The class had only a limited capacity for further growth. Despite this, the FFG-7 class is a robust platform, capable of withstanding considerable damage. This "toughness" was aptly demonstrated when USS Samuel B. Roberts struck a mine and USS Stark was hit by two Exocet missiles. In both cases the ships survived, were repaired and have returned to the fleet.

Through the years ships of this class have proven to be very reliable, multi-purpose vessels that proved themselves in countless operations. They were deployed in a number of conflicts.

Over the past years ships of this class have been extensively modified. In the US Navy the MK-13 launcher disappeared. With it she lost her AAW (anti-air warfare) capability. Leaving her with only a Point defence system (Phalanx). The systems and armament of the remaining ships will be updated as from 2009. The *Perry*-class will be succeeded by the Littoral Combat Ship as from 2019. Other countries also realized modernizations/modifications to lengthen the life of the ships.

Note: Two ships of this class made world news headlines. May 17, 1987 during the Iraq War USS *Stark* was attacked by an Iraqi jet fighter. The ship was hit by two Exocet missiles causing considerable damage and 37 killed.

Under construction at Bath Iron Works in February 1980. Gallery FFG-26 , Jack Williams FFG-24 and Clark FFG-11. *(Photo: MaritimeQuest.com)*

USS Stark listing after being struck by two Iraqi missiles

Less than a year later USS *Samuel B. Roberts* was nearly sunk by an Iranian mine. Nobody was killed, but 10 casualities had to be taken to hospital. For two days the crew fought raging fires and flooding. They succeeded in saving the ship. Two days later the US forces carried out attacks on Iranian oil platforms from which the mines were laid and also attacking merchant ships.
Both units were repaired and recommissioned. USS *Stark* was laid up in 1999 and broken up in 2006.

Technical data

USS *Clark* FFG-11

Class	*Oliver Hazard Perry* (frigate)
Displacement	3.109 tons; 3.993 tons fully loaded; 4.100 tons according to ship's own specification
Length	133,5 m. (445 ft.)*
Beam	13,5 m. (45 ft.)
Draught	7,5 m. (24.6 ft.)
Propulsion	40.000 shp. Two LM 2500 gas turbines (General Electric), one shaft (cp propeller)
Speed	29 knots
Range	4.500 nautical miles at 20 knots 5.400 nautical miles at 16 knots
Complement	200 officers and enlisted

** Long hull measures: 138,1 m. (453 ft.)*

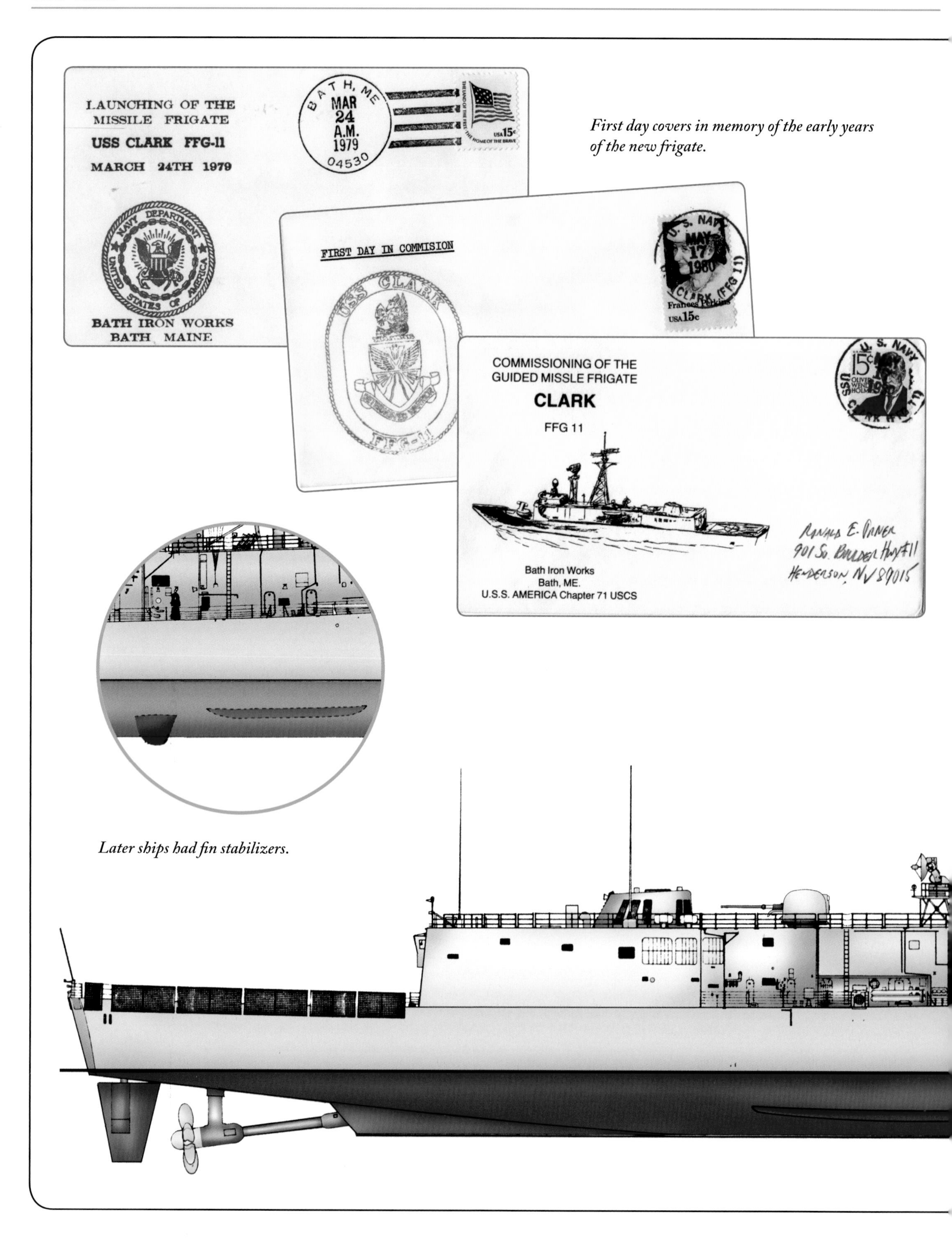

First day covers in memory of the early years of the new frigate.

Later ships had fin stabilizers.

USS Clark FFG-11

Coat of Arms
Shield: Navy blue and gold are the traditional colours of the US Navy and of the Academy of which Admiral Clark was a graduate. The pheons (arrow heads) represent the three wars in which he served. The upper pheon and wings allude to the ship's airborne weapon systems as well as to the Admiral's career-long interest in naval aviation and his skill as a navigator and battle tactician. The pheons further are symbolic of the armament of the FFG-11 and the ability to strike the enemy from a long range.

Crest: The winged sea horse is a symbol of Admiral Clark's career, relating to his close and long association with the air arm in the US Navy and to his foresight and mastery of the combat tactics of air weapon systems. The thirteen stars denote his World War II campaign participation, and the collar bearing the cross represents his highest award, the Navy Cross. The arrowhead refers to his home state of Oklahoma and to his Cherokee Indian heritage; the letter 'C' refers to his last name.

Ship's Motto: "Determined Warrior".

Badge

Cap badge

USS Clark *underway. The FFG was slightly faster than her predecesssor of the Knox-class. The single shaft/propellor was an economy measure allowing the use of a standard* Spruance*-class propulsion cell to be employed. The ship was also equipped with two diesel-driven retractable thrusters that can propel the ship at up to 5 knots, should the main propulsion be lost.*

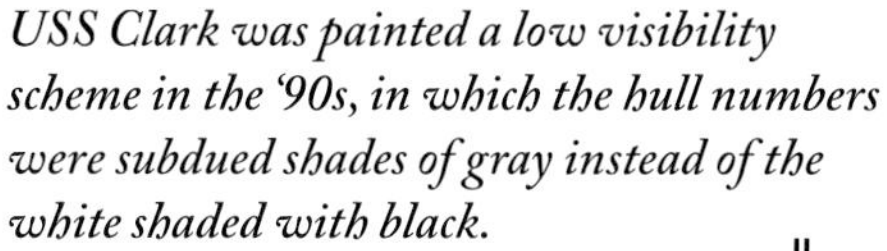

USS Clark was painted a low visibility scheme in the '90s, in which the hull numbers were subdued shades of gray instead of the white shaded with black.

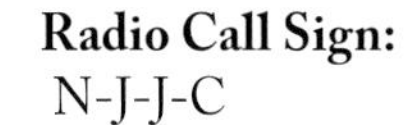

Radio Call Sign:
N-J-J-C

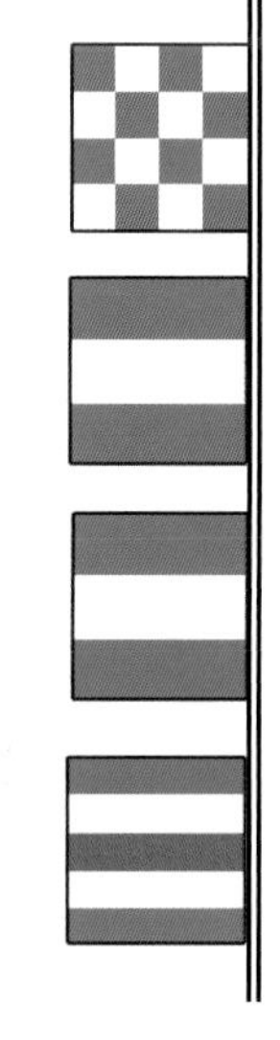

Flag Hoist

Propulsion: Two General Electric LM 2500 gas turbines, 41.000 shp 1 shaft, 1 controllable reverse pitch propeller 5 blades
Two 350 hp Electric Drive Auxiliary Propulsion Units (retractable propeller pods)

Auxiliaries: four 1000 Kw. Ship's service diesel generators

Normal complement:
16 officers, 14 chiefs, 139 enlisted
In addition a detachment of 6 officers helicopter crew and 15 maintenance staff.

Berthed at Port Weller, Canada. Note the absence of a port anchor. Early in their operational lives, ships of this class started to develop serious cracking in the superstructure, which extended from side-to-side and for approximately 70% of the length. These cracks were serious in that they could extend down into the hull portion of the ship providing a way for water to flood important spaces. After detailed inspections and analyses, fixes were carried out on all ships of the class.

Aircraft: SHE-2F LAMPS helicopter

Missiles: 1x MK 13 Mod 4 launcher;
-SSM McDonell-Douglas Harpoon missiles, (launched by MK 13)
- Y36 GDC-Pomona Standard SM-1 MR missiles

Guns: 1x Oto Melara MK 75 3"/62 cal.
1 Vulcan Phalanx CIS 4- .50 cal. MGs

ASW weapons:
6x 324 mm MK 32 (2 triple) tubes / MK 46 torpedoes

Radars: AN/SPS – 49 Air Search Radar
AN/SPS – 55 Surface Search Radar

AN/SPS – 55 Surface Search Radar

AN/SQR-19 Tactas towed array

Vulcan Phalanx

Oto Melara

SPG-60

SLQ-25 Nixie

MK 32 triple torpedo tubes

MK 36 SRBOC Decoy System

Port bridge wing differs from starboard in having a chair mounted. Intended for the commanding officer, but ...

Sonars :
AN/SQS-56
AN/SQR-19 TACTAS (Towed Array)
AN/SQQ – 89 ASW Integration System

Electronic Warfare:
SLQ-25 Nixie
SLQ-32 (V)2 Electronic Countermeasures
MK 36 SRBOC Decoy

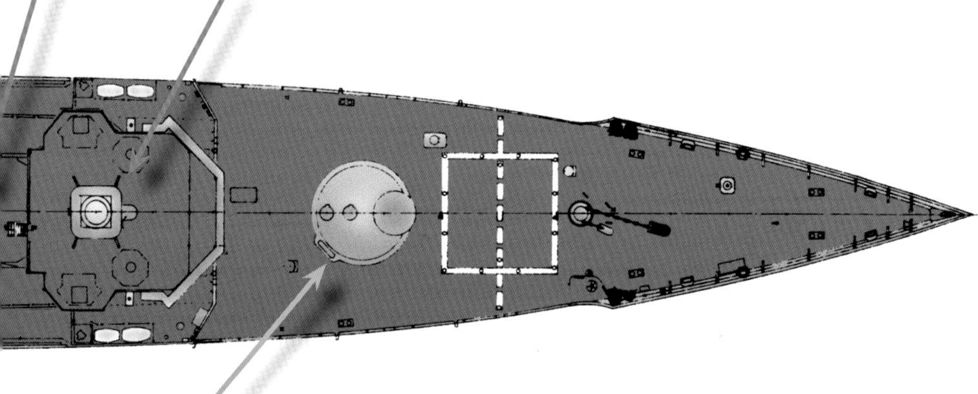

Fire Control:
MK – 92 Fire Control

Weapon-Direction:
FCS 1 modified SPG-60
MK 92 STIR radar (Separate Target Illumination Radar)

Port side in 1997. The boat has been replaced by a RHIB. Behind a new Whiskey-3 UHF SATCOM 'Cartwheel'.

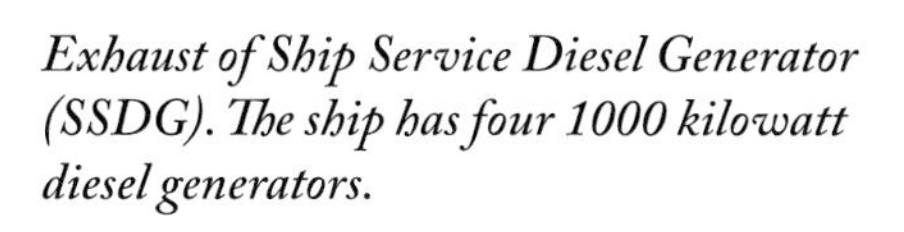

Exhaust of Ship Service Diesel Generator (SSDG). The ship has four 1000 kilowatt diesel generators.

Standard SM-1 MR

The Standard Missile has been part of a system of air defence missiles for 50 years. Initially the SM family was produced by the General Dynamics Corporation Pomona (now Tactical Missiles) Division, which assembled the missile and provided the guidance system. Subsequently Raytheon's Missile Systems Division became the prime contractor and manufacturer. Many of the sub-systems nowadays are dual sourced. The system has been continuously upgraded. The SM-1 and the SM-2 (produced since 1979) are weapon systems for medium range against air targets. They were developed to supersede the Terrier, Talos and Tartar surface-to-air missiles (SAM) of the US Navy. The intention was to produce a new generation of missiles which would fit in existing launching systems.

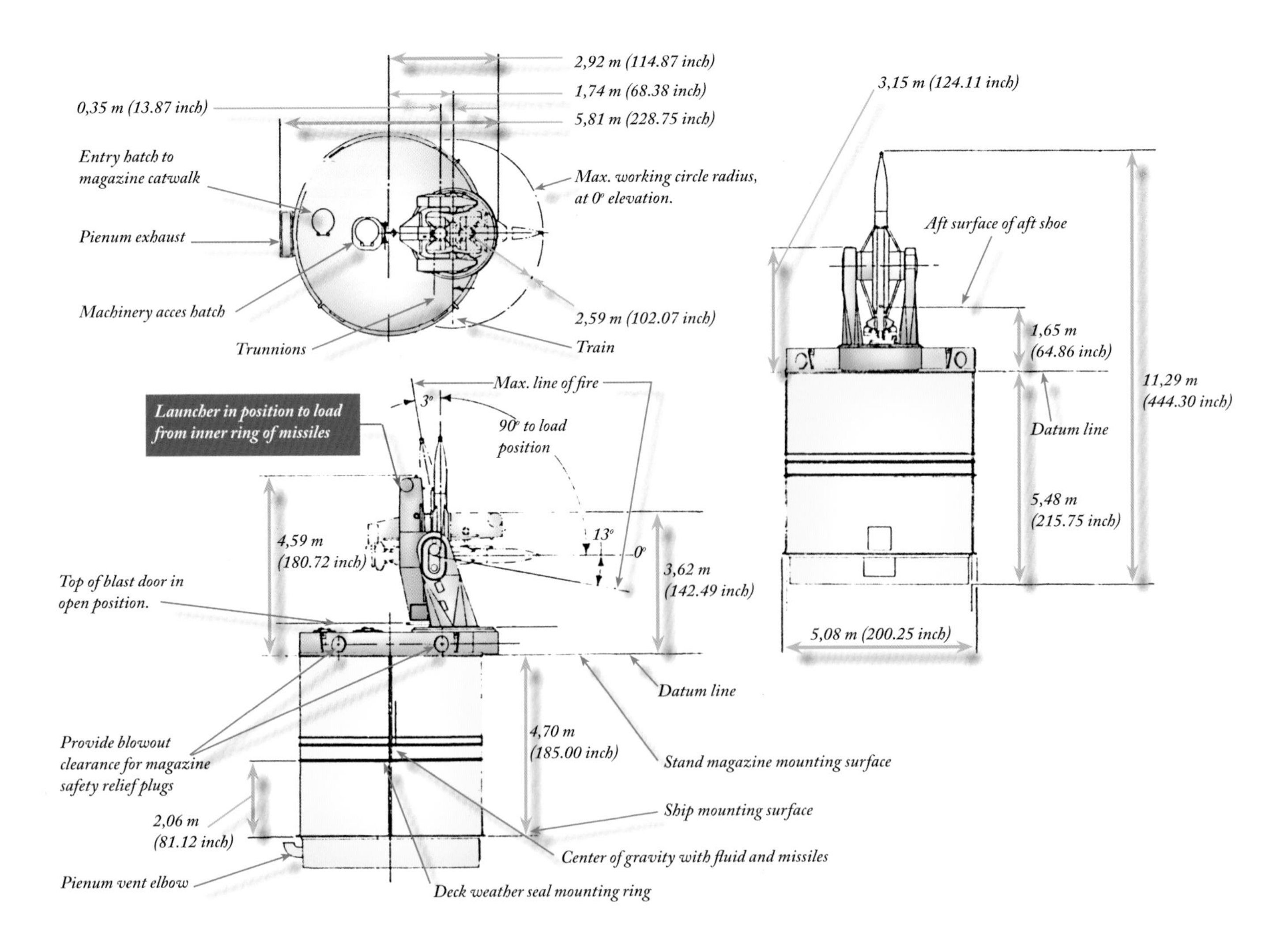

This explains why the new generation of missiles often still carried old names such as Tartar. Originally developed in two versions as medium range (MR) and long range (LR) missiles. (LR with a booster.) Although primarily a surface-to-air missile, there are variants which could be launched against ships.

The latest, the SM-3, has been developed as part of the missile defence system, sea based, Aegis Ballistic Missile Defence System. This system is fitted in the US Aegis cruisers and destroyers for long range intercept. The SM-3 has demonstrated a ballistic missile defence capability. The intention is to intercept ballistic missiles at approximately half-way the trajectory of the incoming missile. Raytheon also develops the warhead for the SM-3.

Disadvantage with the Mk13 launcher is that the single launcher has to launch all 40 missiles. If something happens to that launcher, all missile capability is lost. Second is the nowadays fairly slow rate, only 1 missile every 8 seconds, or a maximum of 7 missiles per minute.

Mid-2000 the US Navy started to remove the frigate's single arm missile launchers and magazines of the *Perry*-class and others, because the primary missile, the Standard SM-1MR became outmoded. One of the reasons to remove this system was due to poor ability to bring down sea-skimming missiles. The other reason was to allow export of more SM-1MR's to allies such as Poland.

Much of the firepower comes from its MK13 launcher and ability to carry 40 medium-sized missiles.

USS Clark

The SM-1 is used against missiles, aircraft and ships. For years it was one of the most reliable in the US Navy's inventory. Here launched during training exercises off the Californian coast.
(Photo: MaritimeQuest.com)

Technical data

Developed in:	United States
Operational:	SM-1: 1968 / SM-2: 1979
Customer navies:	Canada, Germany, Italy, Japan, Netherlands, South Korea, Taiwan, Turkey, and USA.
Specifications SM-1	
Weight:	600 kg.
Length:	4.57 m.
Diameter:	30.50 cm.
Wingspan:	1.07 m.
Range:	74 to170 km.
Maximum altitude:	> 24.400 m.
Warhead:	HE (High Explosive) with impact and proximity fuzes.
Guidance:	Semi-active radar homing.

STANDARD
MISSILE

HARPOON SSM

(Surface-to-Surface Missile)

USS *Clark* was armed with a single, forward mounted MK13 mod 4 missile launcher with 36 SM-1 MR standard missiles and four Harpoon anti-ship missiles available in a below decks magazine (In 2003 the Navy started removing this launcher from existing ships of this class following the retirement of the SM-1 MR missile.)

Technical data

Propulsion	Teledyne Turbojet with solid fuel booster for launching from ships and submarines.
Propulsion power:	660 pounds
Length:	4.75 m (3,84 m without booster)
Diameter:	0.34 m
Wingspan:	0.91 m
Weight	661.5 kg
Speed:	Mach 0.9; 1100 km/hour
Warhead:	Penetrating, lead bearing steel construction, 227 Kg DESTEX (highly active explosive)
Range:	60 nautical miles
Detonation mechanism:	On contact
Remark:	The warhead contains a contact fuse which prevents explosion of the missile prior to launching

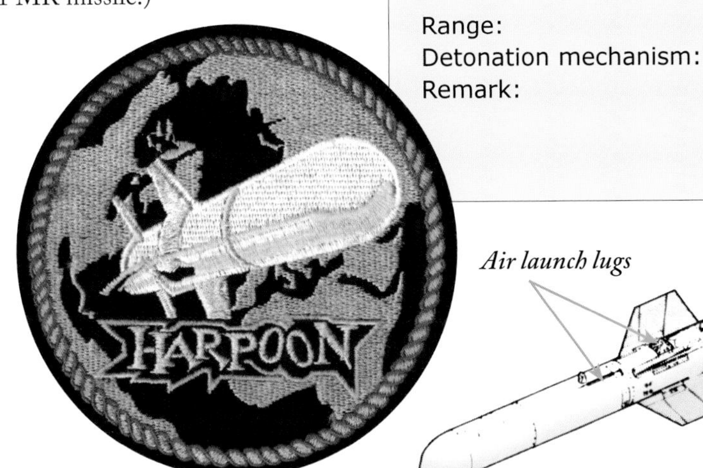

Air launch lugs
Booster
Air-launch
Shoes
ASROC-launch
Aero surfaces extended and folded
Mk13-launch
Shoes
Canister-launch

Harpoon is an all-weather, over the horizon, anti-ship missile system. It originated as a weapon against submarines, as its name implies: a 'harpoon' to attack 'whales'. Starting with a study in 1965 for a long-range anti-ship missile. The Navy project started in 1968 as an air-to-surface missile. Harpoon first flew as a surface-to-air missile on 17 October 1972.

From the beginning, Harpoon development emphasized simplicity and low technical risk. The warhead was originally developed to destroy pressure hulls but then to destroy the compartments which it might penetrate in striking a surface ship.

The US designation for the ship launched SSM is RGM-84. The missile was developed in the seventies of the previous century. Prime contractor Mc Donnell Douglas, later Boeing Defence Space & Security which produced the 7000th missile in 2004. Modifications are indicated by successive Block numbers. The system is continuously updated by improvement programs. The weapon is used on a large scale by the Navy and Air Force of the USA and also in NATO.

It is an active radar searching missile, described as "fire and forget". When entering the target area it activates the homing radar and attacks without any guidance of the launching platform. The advantage of this system is that several missiles can be launched in quick succession without the need for a complicated follow-up and correction system. The Harpoon is powered by an engine propelled by jet fuel. During the flight steering is provided by four electro-mechanical actuators, each for one steering fin. The nose section of the Harpoon is also used for variants of the Tomahawk cruising missiles, which resulted in an anti-ship weapon with a range of approximately 450 km. Currently the Harpoon Block II version is being developed. This provides the Harpoon with littoral-water-anti-ship capability. The Block II will have a greater range and will make use of GPS positioning, enabling the missile to be used against land targets as well.

The helm of USS Thach FFG-43.
(Photo: MaritimeQuest.com)

Crew

The frigates of this class were designed for a complement of: 14 officers; 214 enlisted. In 2010 altered in: 17 officers; 198 enlisted.

At sea a warship will carry out a long list of drills (serials). Independant (e.g. Firex, damage control or emergancy steering) or with other units (e.g. Casex, Seamex or Gunex).

The class was designed under a logistics support concept that emphasized reduced shipboard manning. They have a crew of about 70, fewer than the comparable size frigates of the eighties. The lower manning was attained partly through:

- the use of gas turbine propulsion versus steam power used on previous combatants, and
- the centralization and automation of the control of weaponry and other equipment.

The Combat Information Centre (CIC) in USS Ford FFG-54.
(Photo: MaritimeQuest.com)

The Combat Information Centre (CIC) in USS Reuben James FFG-57. (Photo: MaritimeQuest.com)

The Communication Centre in USS Ford FFG-54.
(Photo: MaritimeQuest.com)

The Communication Centre in USS Gary FFG-51.
(Photo: MaritimeQuest.com)

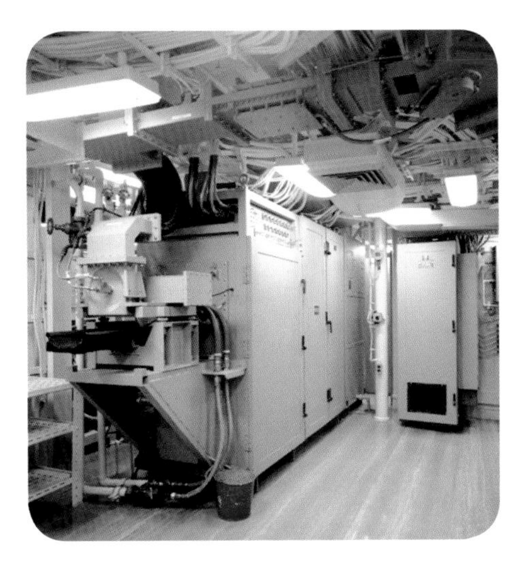

IFF (Identification Friend or Foe) equipment in the CIC of USS Gary FFG-51. (Photo: MaritimeQuest.com)

Antenna

SPS-49

Most eye catching is of course the large SPS-49 antenna.(7.3 m/24 ft wide and 4.3m/14.2 ft high). It is a long range air search radar, unusual in that it is stabilized for roll and pitch. In the long range mode, the AN/SPS-49 can detect fighters at ranges in excess of 225 nautical miles. The narrow beamwidth substantially improves resistance to jamming. There are several versions of this type of radar. It rotates 6 to 12 times a minute.

MK 92 CAS

The Sperry MK 92 Fire Control System is the US version of the Dutch M28. It is a Combined Antenna System (CAS) incorporating two antennas to be used for air and surface tracking. (left) It is a Fire-Control Radar (FCR) designed specifically to provide data for weapon systems.

Port side, with dome of the MK 92. The odd box is SLQ-32, for electronic countermeasures (ECM) it provides warning, identification and bearing of radar-guided missiles and their launch platforms.

Right: The hose connecting probe for replenishment at sea.

The MK 92 fire control system to detect and track air and surface contacts around the ship. The system's radar provides continuously updated information about the contacts to the computer. The computer analyzes the radar data and other information provided by various ship's systems and develops a tactical display. If desired, the computer determines precisely where to aim the MK 75 gun, or programs the Harpoon missile. In addition to weapons control, the system can be used for search and rescue, and navigation.

Mainmast.

The dish on top is the URN-25 Tactical Air Navigation (TACAN) System, a line-of-sight, beacon-type, navigation aid.

What's in the name

AN/SPG-60

Prefix- Joint service designation	1st symbol-installation	2nd symbol-type of equipment	3rd symbol-purpose	4th symbol-series
	A = Aircraft	L = Countermeasures	D = Direction finder	[60th series]
	B = Underwater (submarine)	P = Radar	E = Ejection (chaff)	
	S = Surface ship	Q = Sonar	G = Fire control	
	U = Multi-platform	R = Radio	N = Navigation	
	W = Surface ship and underwater (submarine)	S = Special	Q = Special purpose	
		W = Weapon related	R = Passive detection	
			S = Search	
			W = Weapon control	
			7 = Multi function	

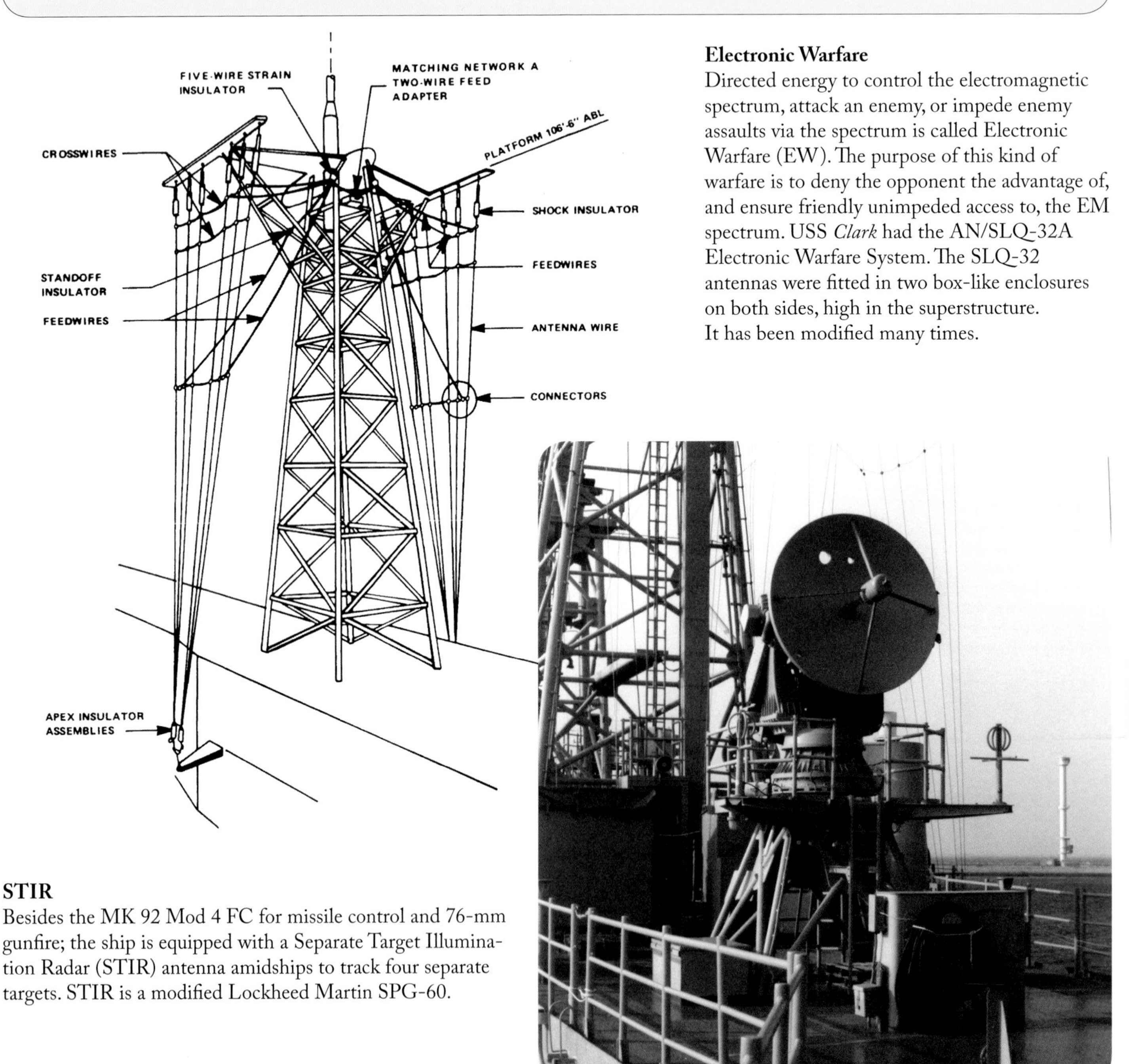

Electronic Warfare

Directed energy to control the electromagnetic spectrum, attack an enemy, or impede enemy assaults via the spectrum is called Electronic Warfare (EW). The purpose of this kind of warfare is to deny the opponent the advantage of, and ensure friendly unimpeded access to, the EM spectrum. USS *Clark* had the AN/SLQ-32A Electronic Warfare System. The SLQ-32 antennas were fitted in two box-like enclosures on both sides, high in the superstructure.
It has been modified many times.

STIR

Besides the MK 92 Mod 4 FC for missile control and 76-mm gunfire; the ship is equipped with a Separate Target Illumination Radar (STIR) antenna amidships to track four separate targets. STIR is a modified Lockheed Martin SPG-60.

WARSHIP

CRUISER HNLMS TROMP — WARSHIP NO. 01

FRIGATE HMS LEANDER — WARSHIP NO. 02

FRIGATE USS CLARK — WARSHIP NO. 04

A new series detailing the full history of the design and history of naval ships. With unpublished photography, drawings/plans of major variants. Detailed information. Warship series are all 48 pages, printed in colour

Lanasta

WARSHIPS ON THE GO ..

- ❑ ***Warship No. 1:*** HNLMS Tromp
  ISBN 978-90-8616-191-1
- ❑ ***Warship No. 2:*** Frigate
  HMS Leander
  ISBN 978-90-8616-192-8
- ❑ ***Warship No. 3:*** Frigate
  HNLMS Jacob van Heemskerck
  ISBN 978-90-8616-193-5
- ❑ ***Warship No. 4:*** Frigate USS Clark
  ISBN 978-90-8616-194-2
- ❑ ***Warship No. 5:*** Protected cruiser Gelderland
  ISBN 978-90-8616-195-9

Softcover | Dimensions 22 x 27 cm | Rich illustrated |
Profile plans | 48 pages | Price: € 14,95

Ordering information

STAMP
BOOKSHOP
HERE

or sent directly to Lanasta,
Slenerbrink 206, 7812 HJ Emmen (NL)

www.lanasta.com

Customer

Name: ....

Adress: ....

Postcode: ....

Place: ....

Country: ....

Colours US Navy

Light Grey	- RAL 7001
Decks	- RAL 9005
Boottop	- RAL 3003

Plans for modelling

The Floating Drydock www.floatingdrydock.com

❶ TFW-FFG16/8: USS Clifton Sprague FFG-16 dated 1986, scale 1/96

❷ G-FFG61: USS Ingraham FFG-61 dated 1995, scale 1/192

Scale Ship Yard offers two sets that includes a glasfibre hull with plans, weapon set and fittings both 1/96 scale www.scaleshipyard.com

❸ WHU-D 2: USS USS Oliver Hazard Perry FFG-7

❹ WHU-D 26: USS McInerney FFG-8

The position of the 76 mm OTO Melara, between mast and funnel, limits its arc of fire.

MK 75 MOUNTING

The OTO Melara 76 mm is a naval gun designed and built by the Italian arms manufacturer OTO Melara. This gun is compact enough to be installed on relatively small naval ships like corvettes and patrol boats. The high rate of fire makes it suitable for anti-missile defence but also against planes and surface targets. It also can be used in the naval gunfire support role (NGS). Custom made ammunition has been developed and available. The gun has been exported in large numbers and operational in 60 different navies.

One gun mount is installed aboard USN frigates and larger USCG cutters (as the MK75). The MK 75 was provisionally approved for operational use in September 1975.

Technical data

Design	OTO Melara
Types	Compact 1963; Super rapid 1985
Manufacturer	FMC Corporation (Ford)
Weight	7,5 ton (17000 lbs)
Lengths	62 calibers
Personnel	Remote control; 4 below decks
Shell	76 x 900 mm; HE 6,296 kg
Elevation	15 degrees / + 85 degrees at 35 degrees per second
Pivoting	360 degrees at 60 degrees per second
Rate of fire	85 rounds a minute
Muzzle velocity	925 m/s / 3035 ft/s
Maximum range	16000 metres
Magazine	80 rounds

BAE Systems (The former Naval Systems Division (NSD) of FMC Corporation) and General Electric Co. (Ordnance Systems Division) were licensed by the gun's designer, OTO Melara of La Spezia, Italy, and competed for the right to manufacture the MK 75 in the United States. In 1975, BAE systems won the competition. One year later licence production was started by the FMC Corporation (Ford) and designated as the MK 75 mounting.

"GUNEX", USS Crommelin (FFG-37). *(Photo: MaritimeQuest.com)*

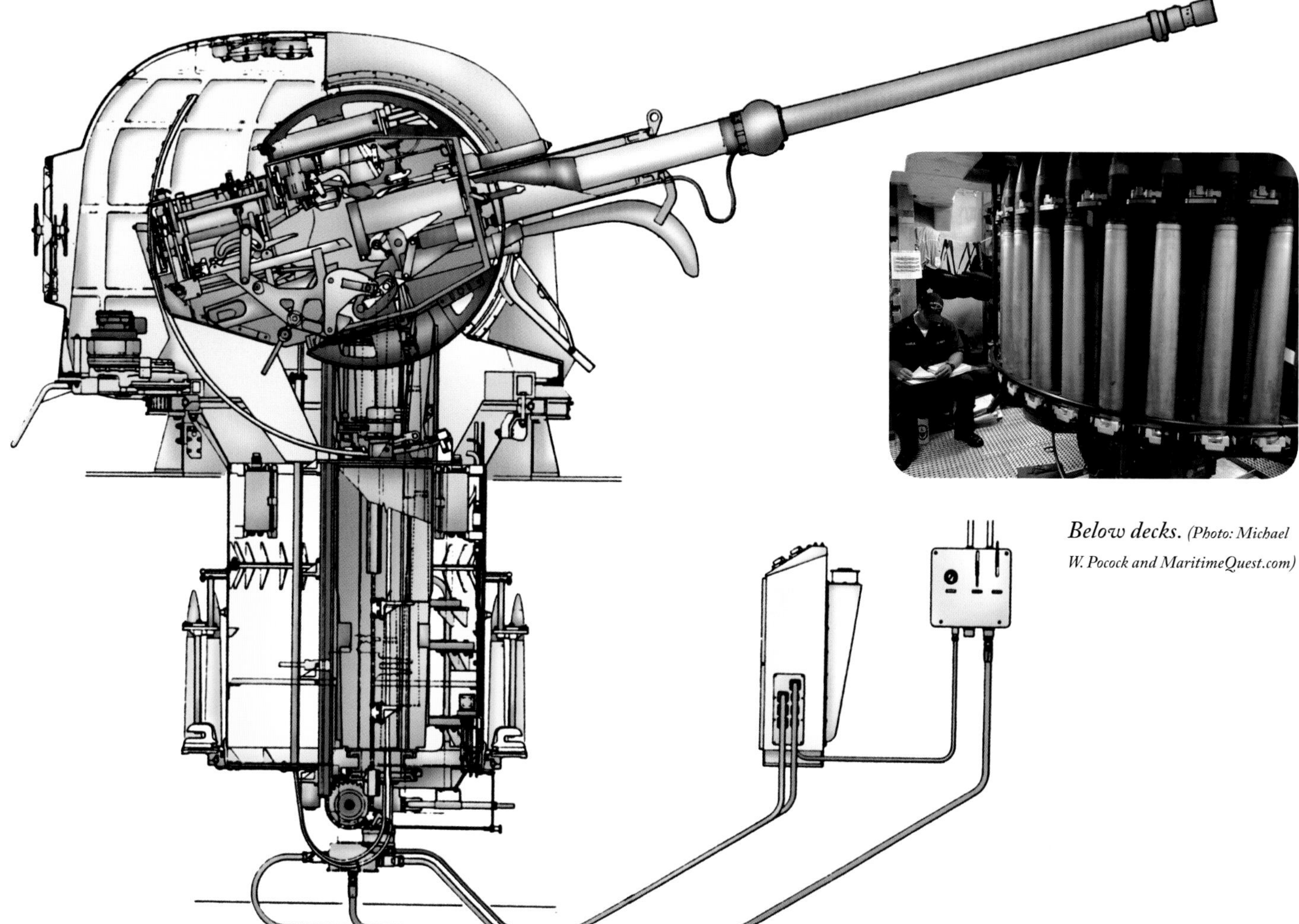

Below decks. (Photo: Michael W. Pocock and MaritimeQuest.com)

The first MK 75 gun produced in the US was delivered in August, 1978. The MK 75 has been modified several times. System improvements include: barrel tube upgrade, breechblock positive stops, and barrel cooling panel upgrade.
Nowadays the US Navy no longer procures the MK 75 but still has logistics support contracts with BAE systems and OTO Melara. (Since late 1994, OTO Breda)

Types of ammunition

- HE (standard on all types) weight 6.296 kg, range 16 km, effective 8 km. (4 km anti-air at 85 degrees elevation.
- PFF anti-missile projectile with proximity fuse and tungsten cubes.
- SAPOM weight 6,35 kg (0,46 kg HE) range 16 km semi-armoured piercing.
- DART guided ammunition for AA targets.

Below decks and forming an integral part of the mounting is the ammunition feed system which consists of a revolving circular magazine around the base. This holds 70 rounds (and 10 in the feed system) which can be fired under remote control from the CIC (operations room). However, for sustained firing four ammunition handlers keep the magazine supplied. Rounds which are vertically stowed are lifted to the gunhouse by a rotating feeder hoist.

TORPEDO MK 46

Torpedoes are intended to be launched against surface ships and submarines. During the Second World War they destroyed millions tons of shipping and caused havoc amongst warships and merchantmen. Today's torpedoes can be divided into lightweight and heavyweight classes; and into straight-running, autonomous homers, and wire-guided. They can be launched from a variety of platforms.

The ship's torpedoes are manufactured by Alliant Techsystems Inc. (commonly abbreviated ATK). This company was formed in 1990 as a subsidiary of Honeywell Inc.

- The MK 46 mod 5 is the standard lightweight torpedo in the US Navy and NATO. The MK 46 was designed for blue water use and introduced in 1960. The torpedo launched from surface warships or aircraft.
  In 1989, a major upgrade program for the Mod 5 began to improve its shallow-water performance, resulting in the Mod 5A and Mod 5A(S). It is expected to be operational till 2015.

Technical data

Mk 46 lightweight torpedo

Design	
Types	Mod 1/6
Manufacturer	Alliant Techsystems Inc. (ATK)
Weight	231 kg (508 lbs)
Length	2,59 m (8.5 ft)
Diameter	32,4 cm (12.75 in)
Warhead	44 kg PBXN-103 high explosive
Operational range	11 km
Maximum depth	< 365 m
Speed	40 knots (74 km/h)
Guiding system	active or passive/active acoustic
Search system	snaking or circling

- The MK 50 torpedo is armed with a 45 kg warhead and has a range of 15 km. The torpedo uses active and passive homing and has a speed of 50 kts. This torpedo was intended to replace the MK 46 as the fleet's lightweight torpedo but it will be replaced with the MK 54 LHT (Raytheon Systems).

Developed to attack sophisticated submarines and can be launched from ships, helicopters and aircraft. This torpedo can also be used in shallow waters. The *Perry's* carry two MK 32 triple-tube (324 mm) launchers. (stbd / port). This type of

The MK 32 triple launcher.

launcher is a standard launcher. Not only in the US Navy but in most NATO navies.
It was manufactured by BAE-systems. There are at least six variants of this launcher. The manufacturer also performs depot-level overhaul, repair, and modernization of all mods for the US Navy and international naval forces.

Launch of a MK 46.
(Photo: MaritimeQuest.com)

The US fleets

The principal operational forces of the US Navy during service of USS *Clark*

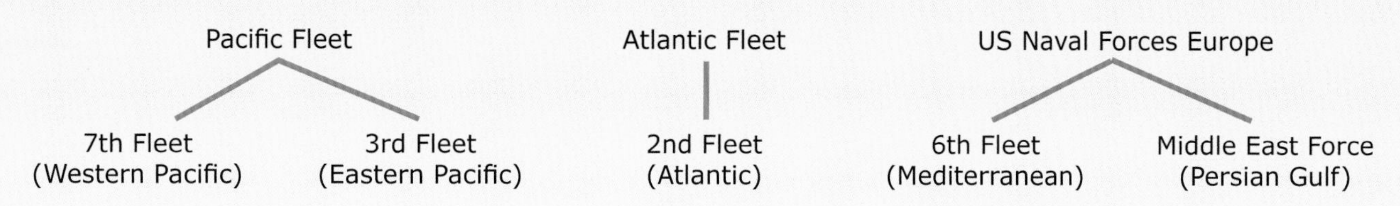

Operational control of these fleets begins with the US President and Secretary of Defense through unified commanders.They have the responsibility for the conduct of military operations.

The fleets tactical organizations are based on Task Forces (TF) based on the relevant numbered fleet designation. When appropriate, subgroupings are designated as Task Groups and Task Units. Thus, TF-62 in the 6th Fleet could have TG-62.1 assigned, this TG can be made up of several TU's e.g. TU-62.1.1, TU-62.1.4.

In general, all major units have operational (TF) designations for use in emergency or contingency situations. Even-numbered ship groups are normally assigned to the Atlantic Fleet and odd-numbered groups to the Pacific Fleet.

Most ships in the forward-deployed 6th and 7th Fleet operate for about six months. Then join the 2nd and 3rd Fleets. While in the latter fleets, the ships receive maintenance, training, exercises and some operational assignments. The 2nd Fleet provides ships and aircraft for NATO in the Atlantic area.

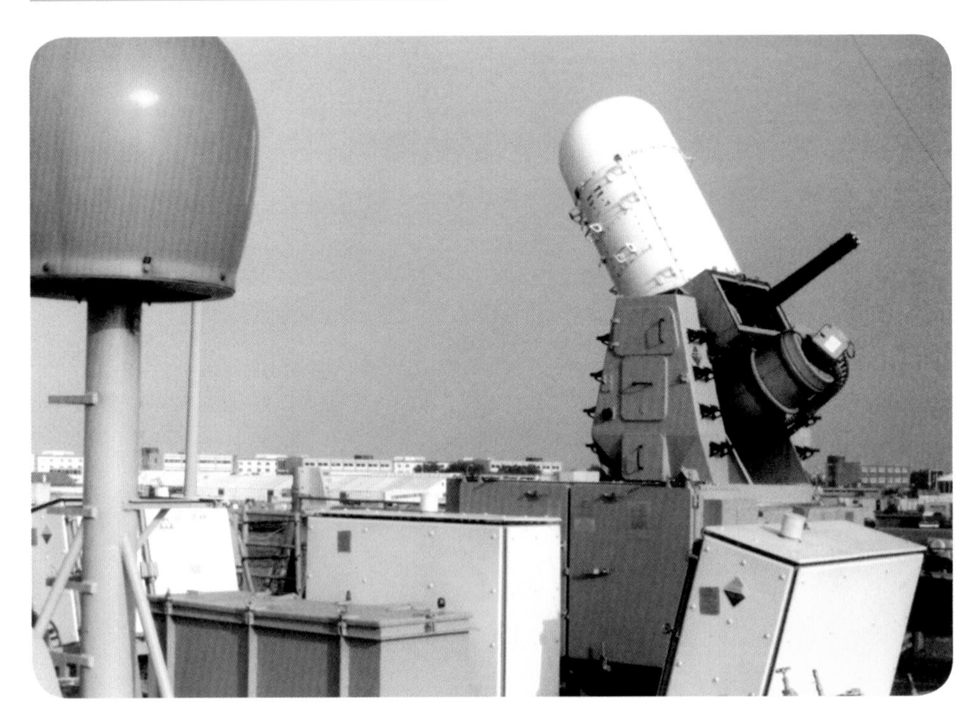

The MK 15 CIWS (Close-In Weapon System) 20 mm Phalanx was fitted late 1988.

PHALANX CLOSE-IN WEAPON SYSTEM (CIWS)

Phalanx CIWS (pronounced "sea-whiz") has been developed to defend against low incoming missiles such as the Exocet. It is the last ditch defence.
The system needs little data from the ship itself. It therefore still functions even when the ship has been heavily damaged. The system only needs a 440 V AC three phase current at 60 Hz and water to cool the electronics.

Basically Phalanx consists of a multi-barrelled Gatling gun with a very high rate of fire and integrated radar fire control. The installation is selfcontained and can be mounted on suitable deck space, like on top of the hangar of USS *Clark*.

The idea of the system is to hit and disable the incoming missile, after which it will veer off course and fall apart.
Phalanx is based on the 20 mm M61 Vulcan Gatling gun that was commonly used in most US Air Force fighters since the 1960ties. As an aircraft weapon it can achieve a firing rate of 6000 rounds per minute. In the naval version, rate of fire is reduced to 3000 rounds per minute. The six barrels are fed from a drum magazine mounted below.

A KU-band radar provides search, detection, acquisition and tracking. While the computer carries out threat evaluation and initiate firing. The whole system can operate completely automatic or in manual mode.

This specially designed frame enables to elevate and train towards incoming targets. It is a stand-alone unit that houses the gun, the automated fire control and all other vital parts. The design makes the weapon suitable for use on ships without advanced systems and sensors.
The first warship to be equipped with a prototype was USS *King* (*Coontz*-class destroyer) in 1973. The trials, modifications and experience resulted in the approval for production in 1978. Production started with 23 units for the US Navy and 14 for other navies. The first ship fully equipped was USS *Coral Sea* in 1980.

A milestone was reached with the production of the 100th system. This was installed in USS *John A. Moore* (FFG-9). In June 1984 frigate USS *Rentz* (FG-46) became the 100th US Navy ship to have the Phalanx installed.

Phalanx Block 1B saw service evaluation in 1999 aboard USS *Underwood* (FFG-36) and was first operationally installed on USS *Taylor* (FFG-50) in September 2000.

Technical data

Phalanx Close-in weapon system

Design	General Dynamics (now Raytheon)
Manufacturer	General Dynamics (now Raytheon)
Weight	12.500 lbs (5700 kg) later systems 13.600 lbs (6200 kg)
Length barrel	Block 0 and 1 (L76 gun barrel) 1520 mm (59.8 inch)
Height	4,7 m (15 ft 5 inch)
Ammunition	Armour piercing tungsten penetrator rounds with discarding sabots
Caliber	20 x 102 mm
Barrel	6 rotating
Elevation	Block 0 -10 / +80 degrees (Block 1 -20/+80 degrees)
Elevation speed	86 degrees per second (both)
Traverse	150 degrees from the center line, speed 100 degrees per second Block 0 en 1
Firing speed	3000 rounds/minute Block 0 en 989 rounds; Block 1 3000 rounds/minute 1550 rounds
Muzzle velocity	1100 m/s (3600 ft/s)
Effective range	3,6 km (1.95 nm)
Gun	M61 Vulcan 6-barrel Gatling
Guiding system	KU-band radar and Infra-red

As with all other weapon systems many modifications were made. Block system 1 (1988) showed improvements in the radar system, the firing speed, increased elevation. Block 1B (1999) added FLIR (Forward looking infrared) systems. USS *Clark* was equipped with the Block 0 system.

Search radar compartment

Tracking radar compartment

M61 Vulcan Gatling gun

Video tracker

Electronics compartment

Ammo storage (1500 rounds)

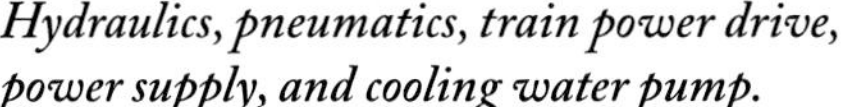

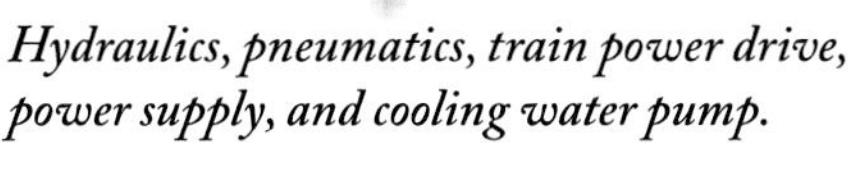

Hydraulics, pneumatics, train power drive, power supply, and cooling water pump.

Kaman SH-2 Seasprite

This helicopter was developed in the late 1950ties to operate from US Navy vessels as a fast multi-purpose helicopter. In the 70's upgraded to an integrated ASW ship based helicopter for submarine attack and OTHT (Over-The-Horizon-Targeting) for ship launched cruisemissiles. The older UH-2 helicopters were rebuilt to SH-2 Seasprite. (Last were taken from active service in 2001.)

Specifications:

Role:	Anti-submarine warfare.
Builder	Kaman aircraft corporation
First flight	1 July 1959 (HU2K-1)
In service	December 1962
Number built	184

UH-2 design was chosen to be the base for the Light Airborne Multi-Purpose System Helicopter (LAMPS) in October 1970. Configuration to SH-2F variant started in the early 70's later also 42 'new' SH2F where built to operate from ships not able to carry the larger SH-60B LAMPS III

The Seasprite is equipped with forward looking Infrared sensors (FLIR), CHAFF flares for radar confusion, dual rear mounted IR scramblers and a missile/mine detection system.

A Kaman SH-2F Seasprite of Light Helicopter Anti-submarine Squadron 94 (HSL-94) on the helicopter deck of USS Clark. Under OPSCON (Operational Control of US Coast Guard) the frigate and embarked LAMPS were deployed in counter-narcotics patrols in an effort stopping drugs trafficking. *(Photo: US Navy)*

SH-2F general specifications

Crew:	3 Pilot, co-pilot/tactical coordinator (TACCO), Sensor Operator (SENSO)
Length:	16.04 m (52 feet 7 inches)
Rotor diameter	13.41 m (44 feet)
Height	4.73 m (15 feet 6 inches)
Weight	3193 kg (7040 lbs.), max 5805 kg (12800 lbs.)
Propulsion	2 General Electric T58-GE-8F turbo shaft 1350 SHP each
Rotor	4 blades main, 4 blades tail
Speed	130 kts (241 km/h) max. 143 kts (241 km/h)
Range	679 km (366 nm)
Service ceiling	6860 m (22.500 ft)
Armament	2 MK46 or MK50 ASW torpedoes

The two hangars are demanding a lot of space..

The FFG-8, 28, 29,32, 33 and 36 thru FFG-61[1] have had their sterns lengthened by 8 feet and the fantail deck lowered by 2 feet to clear the helo landing deck and accommodate the RAST system which required additional space for the helo landing control booth mounted in the deck.

Shipboard landing is assisted through use of a haul-down device that involves attachment of a cable to a probe on the bottom of the helicopter prior to landing. Tension is maintained on the cable as the helicopter descends, assisting the pilot with accurate positioning; once on deck locking beams close on the probe, securing the helicopter to the flight deck. This device, pioneered by the Royal Canadian Navy, was called "Beartrap". The US Navy implementation, based on Beartrap, is called "RAST" (Recovery Assist, Secure and Traverse).

1 *These ships carried SH-60 Seahawk LAMPS III.*

USS Clark was not equipped with RAST.

February 19, 1980: Clark FFG-11 on trials off Maine.
(Photo: MaritimeQuest.com)

***Left:** Compare this photo with the left page and note the modifications to the stern. The openings in the stern are for TACTAS and SLQ-25 Nixie torpedo countermeasures.*

***Right:** The tactical towed array sonar and Bathythermograph room on the USS Kauffman FFG-59.* *(Photo: MaritimeQuest.com)*

TACTAS

The AN/SQR-19 is a passive towed sensor array designed to give surface vessels long-range detection and tracking of both submarines and surface ships. The AN/SQR-19 provides omnidirectional long-range passive detection and classification of submarine threats at 'tactically significant' own ship speeds up to seastate 4, using an array towed on 1700 m cable to provide towed depths down to 365 m.

When excecuting towed array operations the ship will be restricted in manoeuvring and speed. The improved version off the AN/SQR-19A, designated AN/SQR-19B, with four UYK44 computers verses the UYK20 has been supplied from January 1991 onwards. Most have been stored ashore due to the USN de-emphasis on ASW and fiscal constraints. TACTAS operations create a "defended lane"; however some ASW tacticians frowned upon this doctrine.

October 14, 1986: USS Clark underway. *(Photo: MaritimeQuest.com)*

Tow point

Tether cable

Drogue

Hydrophone section

Non-acoustic module

Vibration isolation module

Engineering

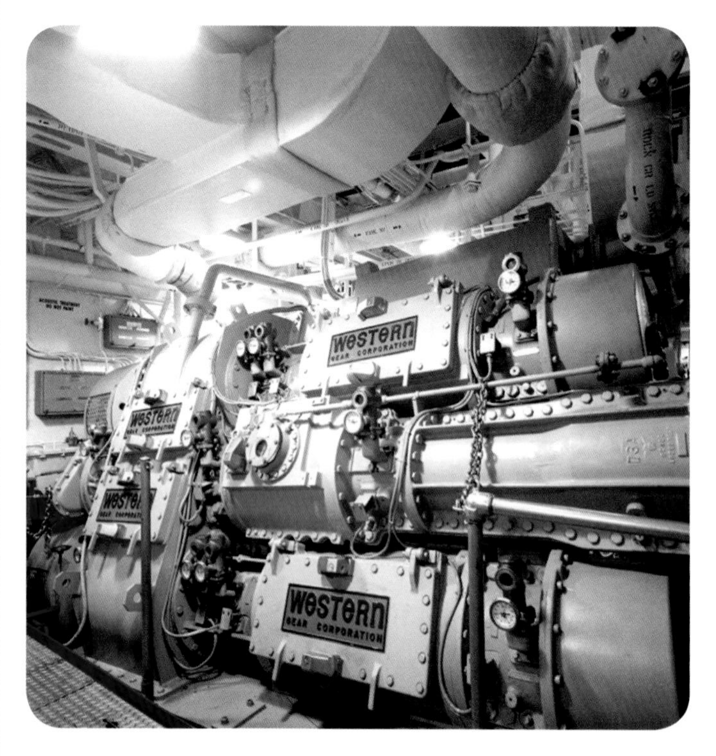

Two General Electric LM 2500 gas turbine engines generating 30.59 MW (41,000 shp) driving a single shaft with a controllable pitch propeller. This single shaft was adopted on cost grounds, but justified on the basis of experience with older escort ships. The thought was that the ship could best deal with torpedoes by burying the engines deep in the ship, that the single shaft itself would be a relatively minor source of vulnerability. In addition, two auxiliary diesel-driven podded retractable propellers of 325 shp each (484kW / 650 shp) for manoeuvring and station keeping. They were designed for a 5 to 6 knots in a calm sea.

The gas turbine control panel in the main engine room in USS Halyburton FFG-40. *(Photo: MaritimeQuest.com)*

The main engine room in USS Ford FFG-54.
(Photo: MaritimeQuest.com)

***Right:** The main engine room in USS Gary FFG-51.*
(Photo: MaritimeQuest.com)

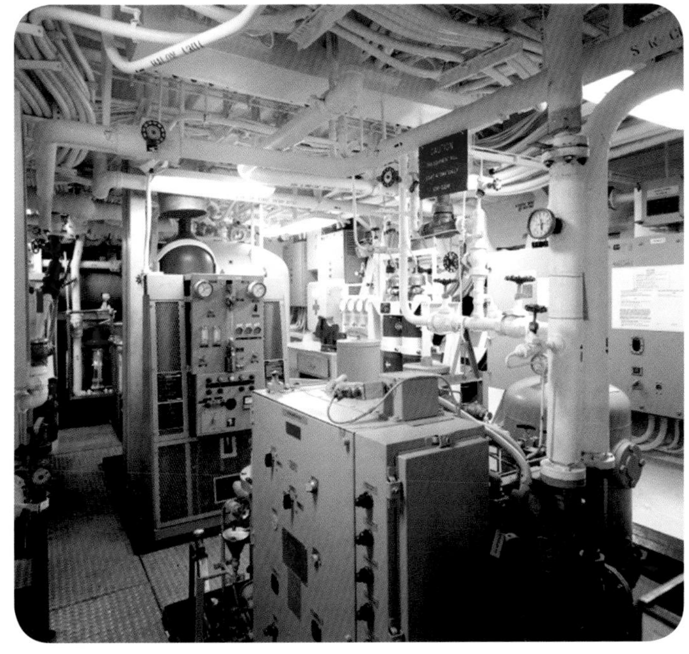

Auxiliary machinery room #3 in USS Ford FFG-54.
(Photo: MaritimeQuest.com)

Auxiliary machinery room #1 in USS Gary FFG-51 (above) and USS Ford FFG-54 (below). (Photo: MaritimeQuest.com)

The LM 2500 was first used in the US *Spruance*- and *Kidd* class destroyers which were built from 1970. This configuration was subsequently used in the *Oliver Hazard Perry* class frigates, and *Ticonderoga* class cruisers. It was also used by one of People's Republic of China's Type 052 *Luhu* class missile destroyers acquired before the embargo.
A LM 2500 unit consists of a gas generator, a power turbine, attached fuel and lube oil pumps, a fuel control and speed governing system, associated inlet and exhaust sections, and lube and scavenge systems, as well as controls and devices for starting and monitoring engine operation.

The propulsion system is a computer-controlled and a derivative of the General Electric CF6 aircraft engine.(used in DC-10 airliner, and Air Force C-5 Galaxy Strategic Transport.)
The plant can be "on-line" and ready in less than 10 minutes, as opposed to the minimum of four hours required for steam power plants.

Chronology USS Clark

A brand new ship, waiting for the ceremony. *(Photo: US Navy)*

1978, July 17	Keel laid at Bath Iron Works, Bath, Maine, USA
1979, March 24	Launched and christened, sponsored by Mrs. Olga Clark, widow of Admiral Joseph J. Clark
1980, February	Sea Trials
1980, May 9	Commissioned, Cmdr. William E. Kelly, USN
1981, September/October	Three weeks voyage in the Caribbean (First of class, to carry out and pass work-up sea training at Guantánamo Bay, Cuba. Receiving outstanding awards.)
1982 January/July	Mediterranean and participating in amphibious exercise Readex'82.
1982 May	Port visit Barcelona
1984, October/May	Mediterranean; Indian Ocean en Persian Gulf.
1985, September 30	Assigned to Naval Reserve Force at Philadelphia
1990,	Recommissioned.
1993, Juni	Pacific en passage through Panama Canal en route Pacific.
1994 /1995	North Atlantic.
1997 April 25,	Returned to Norfolk after a three-and-a-half month deployment in the Caribbean conducting counter-drug operations in support of Joint Inter Agency Task Force, South.
1998 Juni/September	In Caribbean, excercises with units of the Royal Netherlands Navy (UNITAS).
1999 Juni/September	Conduct the Navy's annual Great Lakes Recruiting Tour.
1999 November 15,	Reduced Operating Status pending transfer to Poland.
2000 March 15,	Decommissioned. Stricken from the NVR (Naval Vessel Register); sold to Poland, renamed ORP General Kazimierz Pulaski (a Polish soldier who fought in the American Revolutionary War/ American War of Independence).

May 4, 1989: USS Clark FFG-11 and USS MacDonough DDG-39 berthed at the Manhattan Passenger Terminal for Fleet Week 1990.
(Photo: MaritimeQuest.com)

October 1, 1981: USS Clark FFG-11 returning to her homeport of Mayport, Florida after three weeks training in the Caribbean.

(Photo: MaritimeQuest.com)

Commanding officers

USS Clark FFG 11

May 1980 - August 1982	Commander Wlliam E. Kelly
August 1982 – October 1984	Commander Richard David Williams III
October 1984 - January 87	Commander Arthur Victor Schultz Jr.
January 1987 – March 1989	Commander Wlliam J. Donnelly
March 1989 – March 1991	Commander Victor Howard Ackley
March 1991 – March 1993	Commander Thomas F. Hartrick
March 1993 – September 1994	Commander John Edward Odegaard
September 1994 – December 1995	Commander Bruce J. Cuppett
December 1995 – June 1996	Commander Scott Lee Jones
June 1996 – July 1997	Commander Christopher Charles Cain
July 1997 – March 2000	Commander Albert Francis Lord Jr.

WELCOME ABOARD

UNITED STATES SHIP

CLARK

(FFG-11)

October 22, 1994: USS Clark FFG-11 at Port Everglades, Florida.

(Photo: MaritimeQuest.com)

The Commanding Officers and Crews of
USS CLARK (FFG 11)
and
ORP 272
request the honor of your presence at the
Decommissioning Ceremony of USS CLARK
followed by the ship's transfer to Poland and
Commissioning Ceremony of the Polish Navy Frigate ORP 272
at ten o'clock on Wednesday, the fifteenth of March
Two Thousand
at
Pier 23, Naval Station Norfolk
Norfolk, Virginia

Invitation for the decommissiong.

ORP Generał Kazimierz Pułaski

Clark was renamed on 25 June 2000 in a ceremony attended by Madeleine Albright. The vessel was transferred to Poland along with a second *Oliver Hazard Perry*-class frigate, currently serving in the Polish Navy, as ORP Gen. Tadeusz Kosciuszko. *Generał Kazimierz Pułaski* is homeported at Gdynia and has participated in numerous NATO exercises in the Baltic.

References

- *Combat Fleets of the World (2000-2001), A.D. Baker III*
- *Jane's Underwater Warfare Systems (1999-2000), A.J. Watts*
- *The ships and aircraft of the US Fleet, Norman Polmar*
- *US Destroyers, an illustrated design history, Norman Friedman*
- *World Naval Weapons Systems 1991/92, Norman Friedman USNI*

All photographs by Aad Trapman unless otherwise stated.

Translation
Hans Lok

Author
Rindert van Zinderen Bakker

Consultant
H.Visser (FCCY rtd)

Graphic design
Jantinus Mulder

Publisher
Lanasta

First print, April 2014
ISBN 978-90-8616-194-2
NUR 465

Contact Warship:
Slenerbrink 206, 7812 HJ Emmen
The Netherlands
Tel. 0031 (0)591 618 747
info@lanasta.eu

Lanasta

www.lanasta.com

May 12, 2010: Ships of the Polish Navy in the Baltic Sea.
ORP General Kazimierz Pulaski 272 (flagship) in the centre. (Photo: MaritimeQuest.com)